IMAGINE LIVING HERE

# THIS PLACE IS

# WILD

## EAST AFRICA

### VICKI COBB

ILLUSTRATIONS BY

### BARBARA LAVALLEE

Walker & Company
New York

*Dedicated to Alexandria Nicole Cobb. —V. C.*
*For Hank, who loved a good adventure. —B. L.*

The author and artist gratefully acknowledge the help of the following people: David Herbert, David Jones, and Jennifer Cesnat of African Travel, Inc.; Winnifred Matovu, Marinka Sanc-George, Ben Katumba, Bogere Samuel, and Stephanie Melville of African Pearl Safaris, Uganda; Nampiona Christine, Assistant Education Officer, Uganda Museum; Turyaheebwa Francis of Maryhill High School, Mbarara, Uganda; Miiro Jessica Stacy and Nalcitto Peace, Ugandan Wildlife Club; Jonathan Baranga, Ph.D., Professor of Wildlife, Director, Institute of Tropical Forest Conservation, Uganda; Festo Karwemera and Jotham N. Kanyonyozi, historians, Kabale, Uganda; Tusiime Caleb, Bwindi Impenetrable National Park; Kizito Ssalonglo, Headmaster, New ABC Primary, Kampala, Uganda; David R. Mutekanga, Mwamba Shete, and Aaron Eric T. Shaha, The East African Wild Life Society, Kenya; Simon Ole Makallah, Senior Warden, Kenya Wildlife Service; Hassan Abdi, Luke Abok Otieno, and Samuel Mzee of Mara Safari Club; Masai Chief Ntutu and his son, Stephen Ole Ntutu; Dr. J. H. Grobler, and Derek Potter, Natal Parks Board; naturalist Wayne Beardmore of Kenya; and Dr. Elizabeth I. Macfie, Project Leader, International Gorilla Conservation Programme. Finally, special thanks to our videographer Harland Snodgrass.

First published in the United States of America in 1998
by Walker Publishing Company, Inc.; first paperback edition
published in 2000

Published simultaneously in Canada by Fitzhenry and Whiteside,
Markham, Ontario L3R 4T8

Library of Congress Cataloging-in-Publication Data
Cobb, Vicki.
This place is wild: East Africa/Vicki Cobb; illustrations by Barbara Lavallee.
p. cm.—(Imagine living here)
Includes index.
Summary: Surveys living conditions in East Africa, explaining why
this area is home to the largest and tallest land animals in the world.
ISBN 0-8027-8632-4 (hardcover) —ISBN 0-8027-8633-2 (reinforced)
1. Mammals—Africa, East—Juvenile literature.   [1. Mammals—Africa, East.
2. Zoology—Africa, East.]   I. Lavallee, Barbara, ill.   II. Title.
III. Series: Cobb, Vicki. Imagine living here.
QL731.A356.C635   1998
599'.09676—DC21            97-25454
                                            CIP
                                            AC
ISBN 0-8027-7579-9 (paperback)

Printed in Hong Kong

2  4  6  8  10  9  7  5  3  1

Imagine climbing into a Land Rover and setting out on a safari game drive. You bump off the road over the grassy plains of the Masai Mara of Kenya, East Africa. Yes, you will see animals. Herds of wildebeests, zebras, antelope, gazelles, and buffalo graze on the abundant grass. Hippopotamuses wallow in shallow rivers. With a little luck, you'll see animals that are not as common: lions, cheetahs, elephants, giraffes, hyenas, even a rare rhinoceros or leopard.

The animals are so used to being observed that they don't even blink as tourists take pictures. They are free to roam, but you must stay in your vehicle. Many of these animals are dangerous and will attack if they feel threatened. It's your guide's job to remind you: This place is wild.

East Africa is crowded with more large mammals than any other place in the world. It's home to both the largest and the tallest land animals. It also has the greatest variety of large animals. What is it about this part of the world that led to the development of such extraordinary animal life? What conditions are necessary to maintain the millions of animals that live there?

East Africa is on the equator, where the sun's rays are the strongest and there is no real change in seasons. If you stood on the equator at noon, your shadow would be as small as it can ever be, and a vertical stick wouldn't even have a shadow! You might expect hot weather in East Africa, but since much of the land is high above sea level daytime temperatures are quite pleasant, usually between $75^\circ$F and $85^\circ$F. The region's many different habitats—mountains, forests, and grassy plains, or savannahs—have vast quantities of food and enormous space to roam. These conditions have existed for millions of years. They could support the development of large animals. And so large animals came to be.

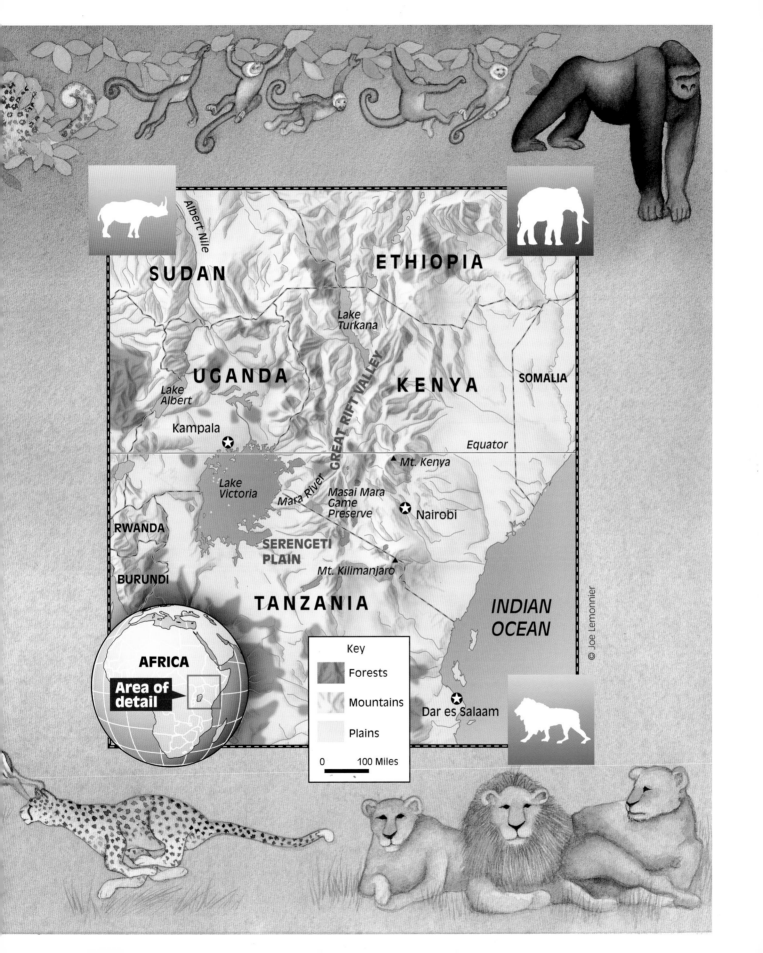

SUDAN

ETHIOPIA

*Albert Nile*

*Lake Turkana*

UGANDA

*Lake Albert*

Kampala ✪

KENYA

SOMALIA

GREAT RIFT VALLEY

Equator

▲ *Mt. Kenya*

*Lake Victoria*

*Mara River*

*Masai Mara Game Preserve*

✪ Nairobi

RWANDA

SERENGETI PLAIN

▲ *Mt. Kilimanjaro*

BURUNDI

TANZANIA

*INDIAN OCEAN*

© Joe Lemonnier

AFRICA

**Area of detail** ➜

Dar es Salaam ✪

**Key**

Forests

Mountains

Plains

0     100 Miles

Elephants are nature's bulldozers. An African bull elephant, at 7½ tons, is the largest land animal in the world. It can walk through patches of trees as easily as an antelope walks through grass. Elephant trails across Africa became roads for people. An elephant's trunk can rip off branches in nonstop sixteen-hour feedings of up to 440 pounds of food a day. Behind them, elephants leave a path of destruction that opens up the forest to new seedlings. Seeds of many plants are spread through elephant dung. Scientists call elephants a "keystone" species. Many other species depend on elephants for their survival, and if elephants should disappear these other species would be endangered as well.

The size of an elephant is its only protection, and it has no natural enemies except human beings. For years, elephants have been hunted for their ivory tusks. In the 1980s, almost 700,000 elephants were killed, more than half the total elephant population. If the killing had continued at that rate, elephants would have become extinct within your lifetime. Fortunately, in 1989, most of the countries in the world agreed to stop buying ivory and to protect the elephants from illegal hunters called "poachers." But poaching is not the only threat to elephant survival. Elephants' habitats are being destroyed by logging and farming. Programs to save the elephants must both catch poachers and teach local farmers how to live side by side with elephants.

At 18 feet, the giraffe is the tallest animal in the world. Its record-length 20-inch tongue wraps around acacia leaves and strips them, thorns and all, into its mouth. A giraffe can run at a steady 31 miles per hour and deliver a nasty kick to a predator.

When it comes to size, mammals are not East Africa's only big creatures. The largest bird in the world—the ostrich—can be as tall as ten feet and weigh up to 300 pounds. Its 3-pound egg is also the largest in the world. One ostrich egg is larger than two dozen chicken eggs. Both the male and female take turns sitting on the eggs, which are laid right on the ground. If an enemy approaches a sitting bird, it lays its neck flat along the ground, looking somewhat like a small termite mound. Reaching speeds of up to 40 miles per hour, an ostrich can outrun most predators. But it won't run and leave its eggs unprotected.

The marabou stork is the largest stork in the world and can be 5 feet long from beak to tail. Like the vulture, it eats dead meat left over from predators' kills. The marabou stork is ugly, yet it is the source of delicate, fluffy feathers that decorate women's clothing.

In East Africa, there is a rainy season from March to May and another one from October to December. The grass grows green after the rains—a delicious treat for about one and a half million wildebeests and more than a half million other grazing animals including zebras, antelope, and buffalo. The animals follow the greening of the grass, traveling thousands of miles from Tanzania to Kenya and back again in a great migration that is one of the wonders of the animal world. When you see them from the air, these large herds of animals look like an army of ants covering the plain.

Wildebeests lead the great migration. They seem to know when it is time to follow the rain. They are accompanied by zebras, who eat a different kind of grass than the wildebeests.

  There are always dangers on the savannah from predators who kill the grass eaters for food. Traveling in herds is one kind of protection as there are many eyes to spot predators. The stripes of a zebra, so striking to the human eye, actually may be a kind of camouflage. As hot air rises from the savannah it creates a horizontal shimmer. Since predators see only in black and white, the combination of shimmering air and vertical stripes creates a confusing pattern. The outline of an individual zebra completely disappears against the background of the herd.

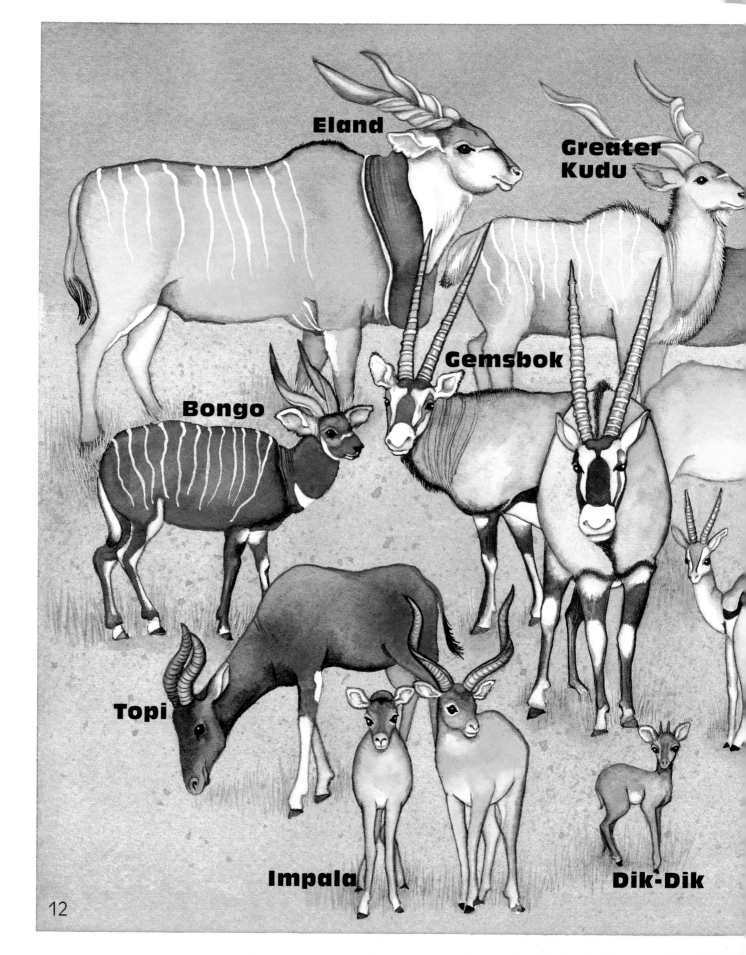

Eland

Greater Kudu

Gemsbok

Bongo

Topi

Impala

Dik-Dik

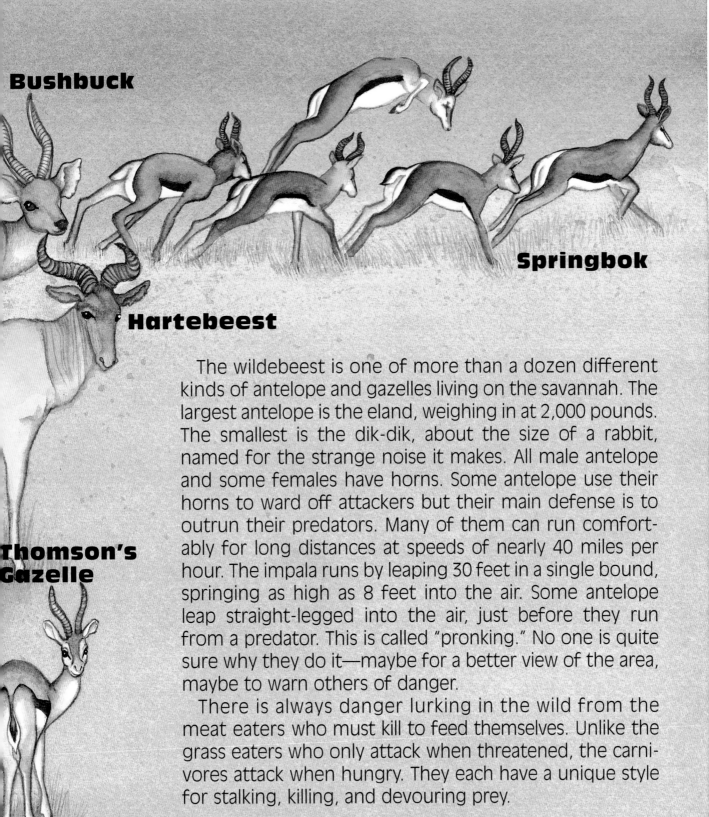

**Bushbuck**

**Springbok**

**Hartebeest**

**Thomson's Gazelle**

The wildebeest is one of more than a dozen different kinds of antelope and gazelles living on the savannah. The largest antelope is the eland, weighing in at 2,000 pounds. The smallest is the dik-dik, about the size of a rabbit, named for the strange noise it makes. All male antelope and some females have horns. Some antelope use their horns to ward off attackers but their main defense is to outrun their predators. Many of them can run comfortably for long distances at speeds of nearly 40 miles per hour. The impala runs by leaping 30 feet in a single bound, springing as high as 8 feet into the air. Some antelope leap straight-legged into the air, just before they run from a predator. This is called "pronking." No one is quite sure why they do it—maybe for a better view of the area, maybe to warn others of danger.

There is always danger lurking in the wild from the meat eaters who must kill to feed themselves. Unlike the grass eaters who only attack when threatened, the carnivores attack when hungry. They each have a unique style for stalking, killing, and devouring prey.

Lions are lazy. They sleep up to sixteen hours a day and hunt mostly at night. They often go hungry because only about one hunt in five is successful. They sneak up on their prey, getting as close as possible before springing to attack. They only can run at full speed for about 40 yards, so it is possible for many animals to outrun them. The females do most of the hunting. After a kill, the male, with his regal mane, gets first shot at the meal, claiming his lion's share. Later, when he's had his fill, the cubs and the females feast.

Cheetahs hunt Thomson's gazelles and many antelope, all fast runners. But cheetahs are the fastest. In fact, they are the fastest animals on land and have been clocked running at 71 miles per hour. Cheetahs approach prey on the run. Unlike lions, they don't have to sneak up close before pouncing. But they can only run fast for short distances. If its prey starts running as soon as it spots the cheetah, it can escape. After a failed attempt, a cheetah must rest for at least a half an hour before trying again.

Leopards look very much like cheetahs except that they don't have the teardrop marking under their eyes. They live alone, mostly in trees, where their spotted coat is almost perfect camouflage. They silently pounce on their victims from the branches.

Some people think hyenas eat only leftover kills of other predators. This is only partly true. They are also excellent hunters and kill most of their food by chasing their prey until it is exhausted. They are noted for their powerful jaws that let them eat not only the flesh of a kill but the bones as well. A group of  hyenas can bring down a wildebeest and reduce it to a skull and a few bones in thirteen minutes!

Life and death may seem cruel in wild East Africa. But there is a balance in nature. The plants support the plant eaters and the plant eaters support the meat eaters in a food chain. Individual animals don't matter on the food chain. It is more important for each type of animal—the species—to survive. Only the strongest animals survive, which is good for the species.

When Europeans colonized Africa in the nineteenth century, it seemed as if there was an endless number of wild animals. Africans killed some animals for food and some in self-defense. But the Europeans killed for sport. At the end of the last century, "safari" meant big game hunting. The goal was to bag one or all of the big five: lion, leopard, elephant, cape buffalo, and rhinoceros. Elephant tusks, rhino horns, leopard and cheetah skins, and the stuffed heads of all kinds of animals became trophies. But several species were pushed close to extinction. As a result, large parks were created to protect the endangered species. In 1977, Kenya banned big game hunting. This helped, but poaching continues to this day. Ivory and rhino-horn trading is illegal but it will also continue as long as there are buyers.

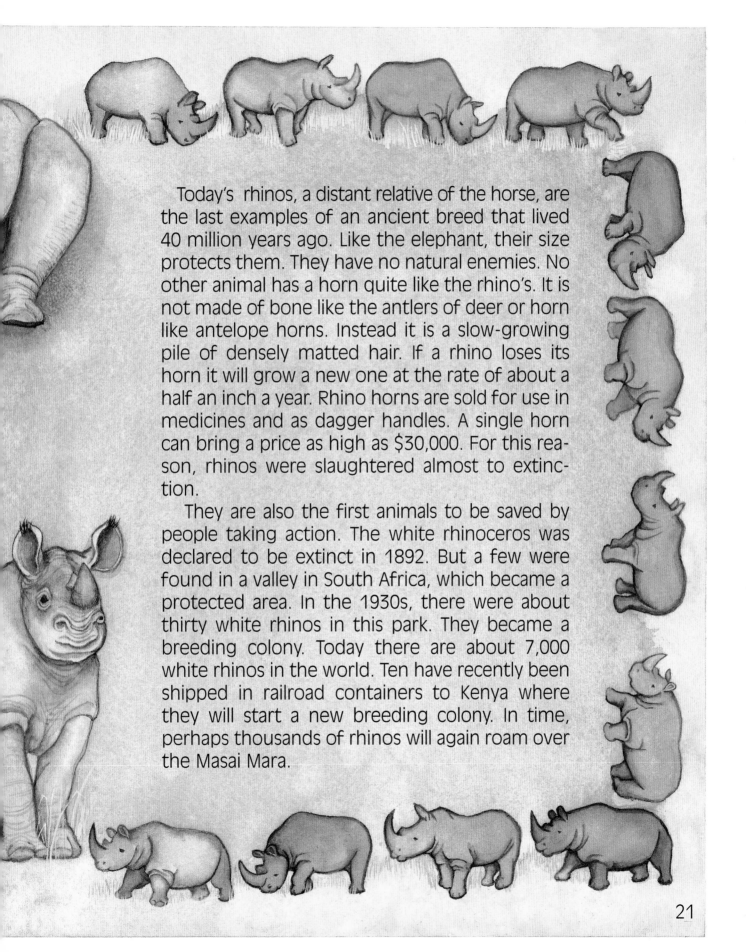

Today's rhinos, a distant relative of the horse, are the last examples of an ancient breed that lived 40 million years ago. Like the elephant, their size protects them. They have no natural enemies. No other animal has a horn quite like the rhino's. It is not made of bone like the antlers of deer or horn like antelope horns. Instead it is a slow-growing pile of densely matted hair. If a rhino loses its horn it will grow a new one at the rate of about a half an inch a year. Rhino horns are sold for use in medicines and as dagger handles. A single horn can bring a price as high as $30,000. For this reason, rhinos were slaughtered almost to extinction.

They are also the first animals to be saved by people taking action. The white rhinoceros was declared to be extinct in 1892. But a few were found in a valley in South Africa, which became a protected area. In the 1930s, there were about thirty white rhinos in this park. They became a breeding colony. Today there are about 7,000 white rhinos in the world. Ten have recently been shipped in railroad containers to Kenya where they will start a new breeding colony. In time, perhaps thousands of rhinos will again roam over the Masai Mara.

East Africa is the home of many primates—the animals that are most like human beings. Leaf-eating monkeys swing rapidly through the trees. Groups of baboons, with doglike faces, groom one another on the grassland. Baboons are the largest of all monkeys. Their size helps to protect them from lions, cheetahs, and leopards.

But the animals closest to us are the great apes: chimpanzees and gorillas. In spite of their size and fearsome appearance, gorillas are gentle vegetarians. Silverback males are the "fathers" of a group. They stand and beat their chests to scare off threats from other animals, including humans, but they almost never attack. They use their great strength to tear apart trees for lunch. As people destroy the forests where gorillas live, they threaten their survival. There are only 650 wild mountain gorillas left on Earth.

Chimpanzees can live in forests and on grasslands because they can eat many kinds of food, including meat. They are not active hunters but will kill a small animal if it happens to cross their path. Chimp babies laugh and cry like humans and even have temper tantrums. Chimps also make tools to catch bees and termites, to crack open nuts, and to soak up water.

It is no surprise that the home of our closest animal relatives is also the home of the earliest humans. Pieces of bone, parts of skulls, and footprints in mud that turned to stone are fossil clues to our ancient ancestors. Scientists, called "paleontologists," found human fossils in several places in East Africa. We know how old the fossils are by the layer of rock they were buried in when they were discovered. The deepest layers are the oldest. The first primates to stand on two feet lived in Africa about 4 million years ago. Modern humans first appeared in Africa about 130,000 years ago and slowly moved north to Europe and Asia. The most outstanding feature of humans over other primates is the size of the brain.

Since early humans walked on two feet instead of four, their hands were free to do other things. Hands could make tools. Tools could be used for hunting, building homes, and making clothing. Hands and tools could be used to make fire. Then people could leave the warm climate of Africa and spread out all over the world. But if we go back far enough, we all originally came from Africa.

One group of people who still live close to the wild animals of the grasslands are the Masai. They live the same way they have lived for centuries. They are a good example of how people can live close to nature without destroying it. The Masai do not hunt wild animals for food. Instead, they raise cows and sheep and goats. One of their main foods is a mixture of milk and cows' blood. They get the blood by opening a vein in a cow's neck. They use the sheep and goats for meat. The Masai women build their homes out of straw and mud. Their huts and fences form a closed circle. During the day the men take the animals out on the savannah to feed. In the evening they bring them home to spend the night inside the safety of the fence. Most Masai only kill a wild animal if it threatens livestock. They think of a lion as a dangerous nuisance.

Today, the Masai have an interest in protecting wild animals—they attract tourists and tourists bring money. Some of the younger Masai are leaving the Masai Mara to go live in cities. Others are farming the land. But those who keep up their traditional ways show visitors how they live and sell them handicrafts.

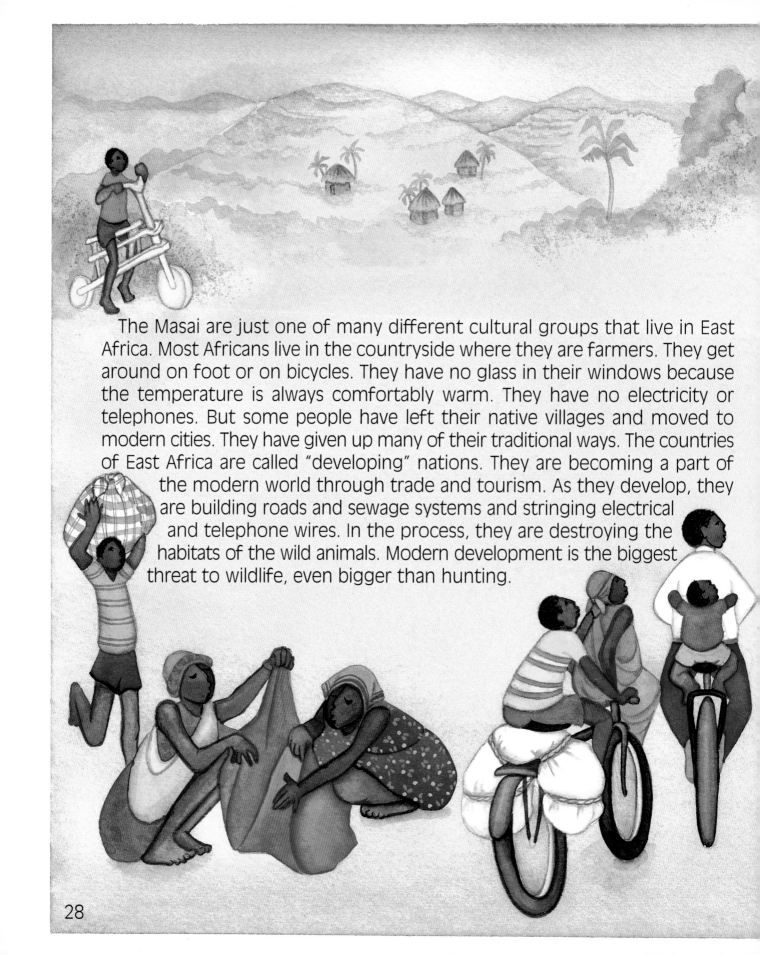

The Masai are just one of many different cultural groups that live in East Africa. Most Africans live in the countryside where they are farmers. They get around on foot or on bicycles. They have no glass in their windows because the temperature is always comfortably warm. They have no electricity or telephones. But some people have left their native villages and moved to modern cities. They have given up many of their traditional ways. The countries of East Africa are called "developing" nations. They are becoming a part of the modern world through trade and tourism. As they develop, they are building roads and sewage systems and stringing electrical and telephone wires. In the process, they are destroying the habitats of the wild animals. Modern development is the biggest threat to wildlife, even bigger than hunting.

What can be done? One thing is to set aside land for animals. Another is to find ways for farmers to share land with animals. Still another is to teach people from all over the world about the incredible African wildlife in its natural setting. There is a new word that conservationists have invented for showing the natural world to tourists. It is "ecotourism."

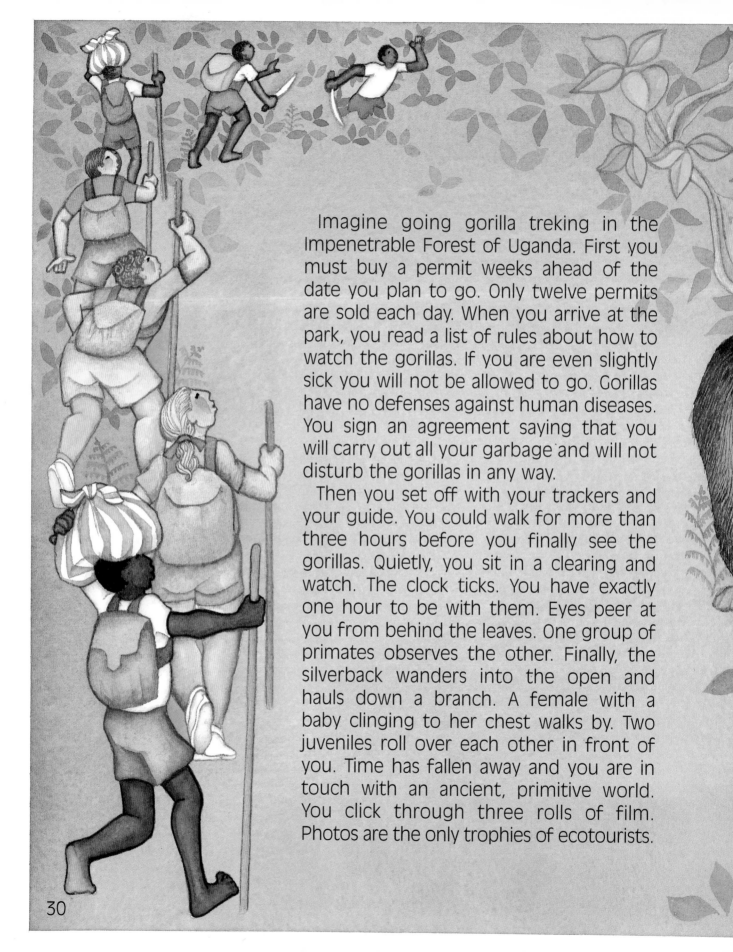

Imagine going gorilla treking in the Impenetrable Forest of Uganda. First you must buy a permit weeks ahead of the date you plan to go. Only twelve permits are sold each day. When you arrive at the park, you read a list of rules about how to watch the gorillas. If you are even slightly sick you will not be allowed to go. Gorillas have no defenses against human diseases. You sign an agreement saying that you will carry out all your garbage and will not disturb the gorillas in any way.

Then you set off with your trackers and your guide. You could walk for more than three hours before you finally see the gorillas. Quietly, you sit in a clearing and watch. The clock ticks. You have exactly one hour to be with them. Eyes peer at you from behind the leaves. One group of primates observes the other. Finally, the silverback wanders into the open and hauls down a branch. A female with a baby clinging to her chest walks by. Two juveniles roll over each other in front of you. Time has fallen away and you are in touch with an ancient, primitive world. You click through three rolls of film. Photos are the only trophies of ecotourists.

The wild animals of East Africa have captured the imagination of the world. They have also captured the imagination of African artists. Masks and carvings and paintings and pottery show images of animals. There are animal stories and animal songs and dances.

People who care about the animals come from all over the world to work with park rangers and in safari hotels. They can imagine living here. Can you?